T E A M

TEAM

TRAINING, EDUCATION, AND MENTORSHIP

DANA R. GORDON

Charleston, SC
www.PalmettoPublishing.com

TEAM

Paperback ISBN: 978-1-63837-739-9
eBook ISBN: 978-1-63837-740-5

TABLE OF CONTENTS

ABOUT THE AUTHOR

CAPTAIN DANA R. GORDON UNITED STATES NAVY, RETIRED

A native of Columbia, South Carolina and graduate of Columbia High School, Captain Dana R. Gordon graduated with a Bachelor of Science degree in Electrical Engineering from Georgia Tech and was commissioned as a Naval officer through the school's NROTC program. He later attended flight school in Pensacola, FL where he learned to fly both Fixed Wing Turbo Prop aircraft and helicopters, earning his Navy Wings of Gold as an Unrestricted Naval Aviator. During his Naval career he also earned his Master's degree in Business Administration from Embry-Riddle University and completed in residence coursework at the Harvard Business School in Cambridge, Massachusetts.

Throughout his stellar 29-year career in the Navy, Captain Gordon completed six deployments on board numerous Navy ships, amassing over six years of total sea duty time, while flying over 3,540 flight hours and seeing 19 countries on five different continents while sailing over eight major oceans and seas throughout the world.

Inspired early in his youth by following the career and achievements of aviation legend and space shuttle astronaut, General Charles F. Bolden Jr., also from Columbia, Captain Gordon worked his way through the rigors of high school, graduating as his senior class president at Columbia High School and was rewarded with a naval ROTC scholarship to Georgia Tech. While in the NROTC program, he earned the opportunity to attend naval flight school and set out on his own path into the world of aviation. After completing the exacting and thorough standards of aviation flight school and obtaining his coveted Wings of Gold as a naval aviator, he received orders to his first fleet helicopter squadron in Jacksonville, FL.

As a young lieutenant during his first fleet squadron tour, while on a deployed mission at sea, he masterfully led his helicopter crew over 100 miles away from their ship through near tropical depression conditions and successfully completed an almost impossible, low visibility, night rescue of five mariners, whose boat had been severely damaged and was sinking after being hit by a water spout. His actions and those of his aircrew earned him his first single action Naval Air Flight Medal and Sikorsky Aviation Rescue pin for lifesaving aerial flight heroics.

Captain Gordon later became a flight instructor where he trained over 280 pilots and aircrew and was named Instructor Pilot of the Year for his squadron and within his Navy Helicopter Air Wing and Region. It was also during this instructor tour that he continued to build on his aviation legacy, earning his second Naval Air Flight Medal and Sikorsky Rescue pin for his participation in a dramatic, two aircraft, overland daytime rescue of six stranded civilians and an aircrewman atop a 670-foot, burning smokestack in Palatka, FL.

Captain Gordon next completed tours as a weapons and tactics officer, officer-in-charge, and department head in multiple squadrons and was named his Squadron and Air Wing's Officer of the Year and the Naval Air Forces and Navy and Marine Association Leadership award winner for his outstanding mentorship and leadership.

He completed a tour at the Pentagon as a member of the Joint Staff and returned to flying as the 20th squadron Commanding Officer at Helicopter Anti-Submarine Squadron Light Four Two (HSL-42) where he was peer selected by his fellow Commanding Officers and was again awarded the Naval Air Forces and Navy and Marine Association Leadership award. While leading HSL-42, his squadron won the Battle Efficiency award, as the "Best in Air Wing Award" winner for an unprecedented three straight years and also won the Secretary of the Navy's Aviation Safety award as one of the Navy's Safest Aviation Squadrons.

Captain Gordon next worked as the Navy's head Aviation Diversity Officer, helping oversee diversity programs and training for over 48,000 personnel in the Naval Aviation community. He was later assigned another tour at the Pentagon as an aide for the Secretary of the Navy, the Honorable Ray Mabus. During this tour he was also named as one of Diversity MBA Magazine's top 100 Under 50 Diverse Executive and Emerging Leaders throughout the United States.

Stepping away from flying, but not leaving the world of aviation, he was selectively chosen by senior naval leadership to command an amphibious assault aircraft carrier, USS IWO JIMA (LHD 7), one of the largest ships in the U.S. Navy's fleet.

As the tenth Commanding Officer to lead IWO JIMA, Captain Gordon guided his crew of 3,250 Sailors and Marines, 31 aircraft, and two amphibious landing boats through an eight-month deployment to the Mediterranean and Red Sea areas of operation. Upon their return from this highly successful deployment, they received the Navy's Battle Efficiency Award as the "Best Ship in its class" and the Secretary of the Navy's Safety and Personnel Retention Awards.

In his final tour of duty with the Navy as the Southeast Regional Operations Director, Captain Gordon led the region's Crisis Action Team through the preparation, response, and evacuation operations of five major hurricanes that impacted 263,000 Sailors and families and over 550 aircraft and ships, across 11 different installations within the Navy's Southeastern United States, Central and South American, and Caribbean Areas of responsibility.

The coordinated efforts of both his regional team, along with FEMA, state, and local agencies helped to ensure that southeastern Naval bases, ranging from Kingsville and Dallas, TX to Beaufort and Charleston, SC along with the Bahamas and Guantanamo Bay, Cuba; their personnel and their military equipment, all remained safe and out of harm's way during these catastrophic events. Captain Gordon and his team's leadership and decision making throughout these events saved the Navy countless lives and billions of dollars because of their work.

After 29 wonderfully fulfilling, eventful, and rewarding years as an officer and aviator in the world's greatest Navy, Captain Gordon and his wife of 28 years, Rolanda Ogletree

Gordon of Dayton, OH, retired in Jacksonville, FL where they now reside. He is currently employed as the director of engineering at Fanatics, Inc., the global leader in licensed sports merchandise and apparel.

Captain Gordon has been a member of Omega Psi Phi Fraternity, Inc, for over 33 years and is also the proud father of two highly competitive soccer players, Danielle and Reese Gordon, who have played and won at the local, regional, national, international, and Division One collegiate level during their careers.

DEDICATION

This book is dedicated to our parents, Bobby and Dolores Gordon and Ron and Laura Ogletree, and our grandparents, A.P. and John Ann Gordon; John and Catherine Martin; Mallalieu Allen and Doris Ogletree; and Jessie and Mary Stanton.

They are the Bridgebuilders who made it possible for us to be where we are today. Without the combined foresight, guidance, and wisdom of each of them leading our family units in their own special way, we as a family would not be where we are today. We love each and every one of you and thank you for all you have done to show us what teamwork, leadership, and, above all else, what love is truly all about.

The Bridge Builder

by Will Allen Dromgoole

An old man going a lone highway,
Came, at the evening cold and gray,
To a chasm vast and deep and wide.
Through which was flowing a sullen tide
The old man crossed in the twilight dim,
The sullen stream had no fear for him;
But he turned when safe on the other side
And built a bridge to span the tide.
"Old man," said a fellow pilgrim near,
"You are wasting your strength with building here;
Your journey will end with the ending day,
You never again will pass this way;
You've crossed the chasm, deep and wide,
Why build this bridge at evening tide?"
The builder lifted his old gray head;
"Good friend, in the path I have come," he said,
"There followed after me today
A youth whose feet must pass this way.
This chasm that has been as naught to me
To that fair-haired youth may a pitfall be.
He, too, must cross in the twilight dim;
Good friend, I am building this bridge for him!"

Dana's Parents

Dolores Gordon, CDR Dana Gordon, Rolanda, Bobby, Danielle and Reese Gordon

Rolanda's Parents

CDR Dana and Rolanda Gordon, Laura and Ron Ogletree, Danielle and Reese Gordon

NOTE TO THE READER

Ultimately the "Key to Life" is finding that one single thing that completes you and makes you whole. When you figure out what it is, you'll have figured out the key to your life.

– Dana R. Gordon

For me, my life has never been about my journey but about what I could ultimately do for others. I've taken on the hard work and struggles I've had to get here, but my road along the way as a Bridge Builder has been about creating the Bridges that I've left behind so others that follow me are able to successfully cross and overcome similar obstacles.

TEAM is a somewhat autobiographical, but mostly historical "lessons learned" book with insight into the things that have thus far helped me in my career as a successful high school student-athlete, college engineering student, naval aviator, ship's captain, and now Director of Engineering at Fanatics, Inc.

My objective here is not to write a "This is how it should be done" or "How to do it" book on Teamwork, but to provide insight into a few of the things I've found that have helped me along the way. It is my hope that you too can find some insight or inspiration through my interpretation of the various quotes, phrases, or words of wisdom that I have heard or taken on-board as a part of my growth and development throughout the various stages of my career.

Thank you for your purchase and for taking the time to delve into the pages I have provided as a part of my journey. I truly hope you enjoy it.

PART ONE

TRAINING

CHAPTER ONE –

Every Day is a Training Day

Long before Denzel Washington reinforced this philosophy in his now infamous movie *Training Day*, and fitness enthusiasts took this motto as a way to ensure we work our bodies into shape, there were Mrs. Bernice Manigo, Pauline M. Davis, Barbara T. Martin, Corine Kimpson, Coaches Hank Betcher, Bobby Young, and Jim Parker, and all the other fabulous teachers on our history and social studies hall at Columbia High School in Columbia, SC.

Every day was indeed a "Training Day" for all their students and student-athletes for four wonderful years, as each of us were taught and trained in more than just U.S. and World History, Government Studies, or Civics. We were given the opportunity to learn many of life's lessons as well as the true essence and meaning of life, along with how to be a responsible and productive citizen from these "old school" teachers and coaches.

This wonderful group of lifelong professional educators were nowhere close to having the demeanor nor did they demonstrate the over the top character portrayed by Denzel in *Training Day*; however, they did take a "tough love" approach towards educating their students to ensure you learned your lessons, both in a book and outside of it, while in their classrooms. You were always accountable for your work and your actions by each and every one of them: no hats on in their hallway, no chewing gum, talking, or sleeping while in their class, and while in their presence you were expected to treat everyone with dignity and respect every single day.

Not that this wasn't the norm for most of our other teachers. I would suspect it is still the expectation with teachers today, but the difference with this group is their education went well beyond their classrooms. The true parental and mentoring influence by each of them and their life skills lessons are the reason many of their students and student-athletes, my classmates and fellow alumni, have all become outstanding and extremely productive citizens in today's society.

I, as well as thousands of other fortunate high school students, had the opportunity of a lifetime to pass through the halls of Columbia High School during the tenure of these fabulous "teachers of history and of life," and we, as well as all our families, our communities, and those we have all gone on to train, educate, and mentor, are so thankful that we did. Thanks to each of you for giving us the solid backgrounds and life-long fundamentals that so many of us still carry with us today.

I'm sure each of you has a "Training Day Mentor," coach, role model, or favorite teacher in your past or in your present life. Be sure you find the time to thank him or her whenever you get the opportunity.

RETIREMENT CELEBRATION

CAPT Dana Gordon with high school teacher and longtime supporter Mrs. Barbara Martin

HSL-42 Change of Command Ceremony

CDR Dana Gordon with high school teacher Mrs. Ellen Parker, longtime supporter and spouse of high school basketball coach Jim Parker

Dana Gordon with legendary Columbia High School basketball coach Bobby Young and lifelong friends Karlton Dixon and John Smith

CHAPTER TWO – YOU GONNA LEARN TODAY

Maybe it ain't the proper way to teach,
but you gonna learn today.

– TIFFANY HADDISH

Think of how stupid the average person is,
then realize that half of them are even stupider than that.

– GEORGE CARLIN

I've always revered great comedians and their art for telling a story that could make me laugh. Tiffany Haddish is indeed one of the great storytellers, along with comedians like Chris Rock, Eddie Murphy, George Carlin, Jerry Seinfeld, Kevin Hart, Richard Pryor, Wanda Sykes, and Dave Chappelle. Besides the obvious side splitting laughter component from each of them, hearing "life's truths" from these comedians is always refreshing and educational in many ways.

I equate their "Truth Telling" and many stories to the opportunity of a lifetime I've had growing up with childhood friends like Royal McKenney, John and Kenny Smith, and Karlton Dixon; college roommates like Robbin Vaughn, Kerry Lavette, Galen George, and Hugo Hodge, and Navy mentors and friends like RADM Earl Gay, RDML Dwight Shepherd, CDR Ken Durbin, and AWRC Bruce Alexander. Of course, I can never forget the invaluable life lessons learned from all of my fraternity brothers of Omega Psi Phi. Listening to each of them and their stories, and more importantly, often times being the butt of their jokes and pranks, I would always learn something that day and come away with a life lesson, whether I liked it or not.

Learning can come in many shapes and forms. I definitely didn't think about it while growing up and living through the trials and tribulations of life, but as I reflect, in many ways, those life lessons from your "truth tellers" are our best and most important teachers. Friendly teasing or "ribbing," as we called it, was often reinforced through laughter, sometimes at your own expense, but were some of the life lessons that you will probably never, ever forget.

Remember, in many instances, getting an education from those life lessons may not be the proper way to learn, but if you're able to see through and look at the truth of the story or event, you're always likely gonna learn something from those experiences.

BEST FRIENDS FOR LIFE

USS IWO JIMA Tiger Cruise and Homecoming Celebration John "Dusty" Smith, James "Royal" McKinney, Captain Dana Gordon, Kenny "Jaws" Smith, and Rolanda Gordon

Retirement Ceremony Celebration

LCDR Scott Palmer, AWRC Bruce Alexander, CAPT Dana Gordon, CDR Ken Durbin, and AVCM Bill Lukander

CHAPTER THREE –

No Process, No Success

Without a Process, there is no Success.

– Dana R. Gordon

Ordinary people focus on the outcome.
Extraordinary people focus on the Process.

– Anthony Moore

Don't focus so much on the wins as the Process. If you have the right Process in place, the winning will come.

– Nick Saban

I have found that there's typically an easy way and a hard way to do things, but that you can get much more done if you develop a process and create an easier way to do it.

– Dana R. Gordon

If I mentioned the names Phil Jackson, John Wooden, Bill Belichick, and Nick Saban, what would be the first thing that would come to mind? With a combined 34 World or National Championships to their credit, the word *winning* or the phrase *winning tradition* would likely be what you would think. Maybe you'd say, they are great coaches and great leaders of their respective teams and players, but I contend that the one thing they all have in common, regardless of personality, leadership traits, or interpersonal skills, is that they each have a proven process or system that they put in place: one that transcends even the greatness of the individual players who play for them and stands the test of time and presents success, almost in a plug and play, next man up, type situation.

I have always found that to be successful at any level, from my most junior positions of leadership to my most senior, you have to have a plan and a process in place from which to deviate. In the absence of a plan, someone else will make one for you or start down a route from which you and your team may not be able to recover. You must always look to develop a process, put it in place, and at least have something to steer your team towards; otherwise just like a rudderless ship or a tail rotorless helicopter, you will eventually drift or spin out of control and crash.

Coaches coach, and leaders lead, allowing players and employees to execute, but proven processes are always successful and are what ultimately win championships. By making your team focus on their daily mission and teaching them the correct, winning processes you want, you can get them to focus on just doing their job within that winning process, which will eventually lead to success.

Within the two organizations I had the pleasure of leading during my time in the Navy, I always found that once we had the right proven processes in place, we then just needed to ensure those carrying out their daily duties did the job they were supposed to do. "Just do your job" wasn't a catch phrase, but had relevant meaning since as we know, "a chain is only as strong as its weakest link." This theory is never more clearly proven once you have the correct process in place. Once your plan or process is implemented, as a leader or coach you should be able to consistently identify what you need to improve upon to make your team that much stronger, make those adjustments, then drive your team towards the ultimate success you are trying to achieve.

CHAPTER FOUR –

TRAIN HARD, PLAY HARD

You train the way you play.

– MICHAEL JORDAN

I make my practices real hard because if a player is a quitter, I want him to quit in practice, not in a game.

– PAUL "BEAR" BRYANT

If you want to win or be successful in anything you do, you need to train and study just like you're doing the actual event. You have to "train the same way you expect to play." You never know when your number may be called to perform a mission on short or no notice, but if called upon, by preparing ahead of time, your ability to perform that task becomes that much easier.

Throughout my many years flying as a helicopter pilot in the Navy, we trained on every flight we went out on, as it was an absolute requirement. Yes, every day was indeed a "training day" for us if we weren't on an actual deployed, real world mission, and even then we still had minimum training and qualifications that we as aircrew always had to perform to stay proficient. Those thousands of hours of training, flying both in the simulator and aircraft, are always there for those "just in case" or "oh shit" moments you may have when your number is called, and you are required to be the best version of yourself.

Those hours of training as a pilot indeed helped me safely and successfully complete 3,540 mishap-free flight hours, over 3,000 safe landings both ashore and onboard ships at sea, along with two very difficult, non-standard lifesaving rescues which saved the lives of six people. That training indeed tested my piloting and teamwork skills and also came in handy with numerous unexpected aircraft mechanical failures or emergency situations that you always had to be prepared for since there is no ejection seat in a helicopter.

How important is it to be trained and ready? In my at-sea rescue, I was awakened by my Officer in Charge at 0345, told what was happening, was briefed by the shipboard team on the situation and stranded boat's general location, briefed my crew and we launched an already prepared and spinning aircraft in less than an hour from me being sound asleep.

During my overland rescue, my crew and I were the senior members helping coordinate the logistics with the primary Search and Rescue aircrew when we were asked 30 minutes later by our Commanding Officer to join them on scene, "just in case." Although not expecting to be a part of the rescue, we briefed and prepared as though we would do the mission as a rescue unit, because it's what you do to *always be ready*. Ultimately when the other aircraft had an engine failure due to smoke ingestion, we found ourselves a willing participant heading towards the same 670 foot tower engulfed in smoke to help rescue one of our own. With our rescue aircrewman now stranded, on his knees trying to take in what little air he had left in his emergency breathing device, we were able to swoop in as quickly as we could to get him.

Training as a team on a larger scale is also just as important during a dangerous evolution when the mission is critical and lives are at stake, and this was never more evident to me during my shipboard time as Executive and then Commanding Officer onboard USS IWO JIMA (LHD 7). Shipboard handling and maneuvering, underway replenishments, damage control exercises, steam plant engineering drills, man overboard drills, flight deck aircraft emergency landing, well deck and small boat evolutions were all part of the constant training that was an absolute necessity for both me and my Sailors to ensure we performed at our very best.

At some point during the day and especially at night, each of those skills and proficiency levels were tested by us all and were vital in determining if we were successful in conducting our mission. Fortunately for all of us, during my three years onboard, we performed well enough to not have any injuries, loss of life, significant damage to aircraft, watercraft, or shipboard equipment, and not be the next front page article on the *Navy Times*, highlighting our failures.

While those numerous evolutions were definitely intense, the amazing ballet-like proficiency and orchestration of each of those operations would never have been possible without the thousands of hours of combined training by all 3,250 Sailors and Marines that I had onboard who believed in the process, trained to it daily, and did their specific jobs very well. In every evolution, we trained the same way we expected to do the exercise or carry out the mission. If we didn't do the training well, we stopped and tried it again, time and weather permitting, until we got it right. Why? Because in every instance, whether it was flying an aircraft for me or onboard a ship with my crew, my life or the lives of others depended on us being successful.

So how does this apply to the overwhelming majority of you, the readers, who will never be faced with a life or death situation and need to be as close to perfect as possible like we had to be? Well, if you're a coach, a director, supervisor, or manager, or someone who is responsible for getting the most out of a team or group, why would you not do what you could to ensure your team trains the way you would want them to perform when it matters? If you personally have a speech or presentation to give, why would you not train and practice either of them until you have your timing and delivery down the way you expect it to happen? Anything short of training and working towards that delivered excellence when it matters, and I would say you have set yourself and/or your team up for failure.

As a soccer coach, why would you not practice penalty kicks before a tournament if you know it might come down to your players making those PKs to win a game? In basketball, why wouldn't you shoot your team free throws at the end of practice, just like at the end of a game when your players are tired? In flight school we used to do what we call "chair flying" where we would sit in a chair, just like we were in a cockpit, and mimic the actions, hand movements, and all radio calls from takeoff to landing and visualize a complete flight. Why? Because in an aircraft, time gets compressed, especially in some cases when you're going several hundred miles per hour, and that is not the time you want to be taking or going through those actions mentally and physically for the first time.

Today I still use this same technique prior to any speech or presentation I'm going to give to ensure I'm ready, which always allows me the ability to adjust to changes or obstacles that may come my way since at least I've practiced it several times. "If you stay ready, you ain't got to get ready, since success happens when opportunity meets preparation," so be sure you train the same way you expect to play.

It all sounds pretty simple on the surface, but you'd be surprised how many people fail to do this. Then when they revisit or debrief why they failed, this is the number one thing they have neglected to do in their preparation process. Don't shortchange yourself or your team by not preparing and training for what you want the desired outcome to be. If you do, in the end, if you fail or are not as successful, you have no one to blame but yourself.

Yep, there's a Helo flying inside all that smoke
HSL-40 rescue of five stranded civilians and aircrewman on top of smokestack in Palatka, FL
Aircrew: LT Billy Carter, LT Dana Gordon, AW2 Eric Kazmerchak and AW2 Scott West

Danielle Gordon taking a shot during her teams National Championship run

Reese Gordon taking aim on goal during a high school match

CHAPTER FIVE – IT DOESN'T HAVE TO HURT

I'm a Sailor, sir. I can fix stupid, but it's gonna hurt.

– MASTER CHIEF PETTY OFFICER MARCUS "CADILLAC" LEALIE

Surround yourself with people who are smarter than you.

– STEPHEN ROSS, OWNER MIAMI DOLPHINS

While onboard the ship one day when discussing solving a problem for me, I got these now infamous words of wisdom from one of the best and most respected Master Chief Petty Officers I've ever had the pleasure of serving with – "Cadillac" Lealie. His words were so simple, yet so profound, that as a leader I had to take a step back and fully grasp the true meaning of this precautionary warning. I needed to fully appreciate and also know why I should take heed and truly understand the ramifications of what may happen when asking to get something done. Before you set out to solve a problem, be sure you look at what the unintended consequences of those actions might be, and time permitting, always ask those involved for feedback on your course of action.

In positions of leadership it's always best to surround yourself with subject matter experts, people who are smarter than you are. You should always ensure you have someone who can give you some perspective as well as possible outcomes (good or bad) on decisions you are looking to make. As a leader, we must always be mindful of tasks that we ask of those who work for us to accomplish. Without having the appropriate experience or experienced personnel to bounce those decisions off of to help guide us, we may be setting ourselves, and ultimately our teams, up for failure. The conversation with my Master Chief was one where, after listening to all the second- and even third-order effects of those initial actions to "fix" things, we came to a better, more thought out solution to solve the problem... and in the end, we got it done the right way that benefitted all, and it *didn't hurt*.

CAPT Dana Gordon and ITCM Marcus "Cadillac" Lealie

CHAPTER SIX – TAKE THE SHOT

You miss 100 percent of the shots you don't take.

– WAYNE GRETZKY

Wise words from *The Great One*, but oh so true. Whether taking a shot in sports, or making a decision as a leader in a corporate setting, you must take action or you will ultimately fail.

Another great saying that could accompany this is, "Indecision is a decision within itself," as you can't succeed unless you make an attempt to do something when given an option, so you therefore have to take action to continue moving you or your organization forward. There are many who will look for the "perfect solution" which will never come and be paralyzed with fear of failure, all the while allowing time and opportunity to pass them by.

If there is no obvious correct path or direction to follow in order to find a solution, then perhaps you are trying to solve a Wicked Problem, one that may have incomplete, contradictory, and changing requirements for which there is no single solution. When faced with a Wicked Problem, I've found the best option may be to try and dissect the problem components individually and then work the overall solution by tackling them one variable at a time.

No matter what though, you must take a shot; you must take action towards trying to find a solution. Otherwise you or your organization will be doomed to fail under the weight of Paralysis by Analysis, as you over-analyze things while trying to find a solution to your problem.

CHAPTER SEVEN – Training, Preparation and Winning

If I had six hours to chop down a tree,
I'd spend the first four sharpening the axe.

– Abraham Lincoln

It's not what we do once in a while that shapes our lives;
it's what we do consistently.

– Tony Robbins

The key is not the will to win; everybody has that.
It is the will to prepare to win that is important.

– Bobby Knight

Only place *success* comes before *work* is in the dictionary.

– Vince Lombardi

There are many, many more quotes I could have used here on winning and preparation. I figured I'd give you a few from some of the best in the business at what they did. I think individually they each speak for themselves and are all a part of the larger explanation of why planning, preparing, and training for any event, game, or process you are looking to undertake is absolutely key to you being successful. Sharpening your tools, honing your skills, or making sure you have what is necessary to be successful is always important. You'll always want to give yourself the advantage by going into any battle with the best tools and skills you have available to you, be it on the field of play, in a foreign territory, or in a boardroom. Never bring a knife to a gun fight.

Consistent training and honing of those skills daily, sometimes hourly, is always the key to making sure you're the absolute best at what you do. Malcom Gladwell presented a case in one of my favorite books, *Outliers,* where he talks about the 10,000 Hour Rule, asserting that the key to being the absolute best at anything can be achieved by practicing it correctly for at least 10,000 hours. While there may not be a set number of hours in specific occupations to support this, I am a firm believer that there has to be some real legitimacy to this as 10,000 correctly practiced and performed hours is a lot of time on task for any given occupation.

Practice does make perfect, but more importantly, *perfect practice increases your level of perfection*. Lots of words, but the bottom line is that while you may have the will, the work ethic, and the determination to do things, it is just as important to do them correctly that will ultimately make you successful. No one is a "natural," or becomes great without persistent and continual preparation. Those who are truly great, the *G.O.A.T.s,* have the rare ability to combine great skill and the will, along with an unnaturally high work rate and work ethic, to become successful.

If you strive to be the best at what you do, then sharpening your tools, preparing yourself mentally and physically for your task, and training consistently will be key and are a few of the necessary skills to make you competitive in your field. To overachieve and move ahead of all others and reach the next level of success, I believe you must also couple these with the will and a relentless work ethic necessary to surpass all the others.

CHAPTER EIGHT –

Give What the Group Needs

You have to give what the group needs,
instead of what you want to give.

– Kara Lawson

To be a great coach, you can't be afraid to get fired.

– Patrick Mouratoglou,
Serena Williams' coach

Just about everything you do in life revolves around some aspect of being part of a team. Whether you are in a personal relationship, a member of a sports team, part of a team or group at work, or a member of what I always called the greatest team sport in the world as a member of the United States Military, everything revolves around a team and a team dynamic. Even within individual sports, there is some form of team aspect, as the greatest individual athletes like Tiger Woods and Serena Williams have a team of supporters, coaches, and trainers to make sure they train consistently so they are able to play at the very top of their game.

So what makes for a successful team and team dynamic? I think the most profound words I've heard on this subject were provided by Duke University women's head basketball coach, Kara Lawson, when she talked about how as a player you must make the personal sacrifice of yourself to, "Give what the group or team needs to be successful, instead of giving what you individually want to give."

We've all seen some teams composed of great individual stars who on paper were "can't miss champions," but somehow once they came together, some fell short of attaining any level of success, while others failed miserably in their quest to be the very best. So why is that? I truly believe it's because unless you can convince all members of your team to give some part of themselves for the betterment of the team's success, then you will never truly get the highest, most successful performance from your team, group, or organization.

As a point guard in basketball, I was always happiest when I made the extra pass and watched my teammates make an easy basket, take a momentum shifting jump shot, or make a crowd-thundering dunk. One of my favorite players of all time is Earvin "Magic" Johnson because he made the art of passing fun. The "magic" of how he did it was what provided the *ohhs* and *ahhs* from the fans, and as a player, those moments always provided an adrenaline rush when you could make a pinpoint or No-Look pass that led to a bucket. The willingness to "give up the rock," even in some cases when you may have a shot, to ensure someone else has an easier opportunity is the true definition of giving what the group or team needs, not what you want.

At Fanatics, we have a term we call #ONEFANATICS, and I'm a firm believer that this philosophy is framed around those same selfless principles. We have numerous teams throughout the country and the world working together daily. Without someone making the sacrifice to give of themselves or their group, for the betterment of the team, I don't think we would have as much success as we've seen to date within the licensed e-commerce sports merchandise industry.

As a group leader or coach of a team, you can't be afraid to ask those who work for you to make those sacrifices. The coach who lets his star player "get away" with things will soon lose team cohesiveness because in many cases they are not asking that player to make a sacrifice for the betterment of the group or team. I've always found that when the leader holds everyone accountable, including him or herself, that the team trust and cohesiveness is raised to another level, a Championship Winning level, that once engrained in those who are a part of it, becomes the expectation and the norm of the group.

I've proudly been a part of some of the most successful organizations within the military and the civilian world and also had an opportunity to see great teams and groups from afar. The common thread to all of them was this "self-sacrifice" for the betterment of the team. Without these actions, I truly believe you will never attain the true goals you may want to achieve for yourself or those of your team.

The United States Military, the Greatest Team Sport in the World

USS IWO JIMA Underway on deployment in the Red Sea with one third of her Sailors and Marines of the 24th Marine Expeditionary Unit assembled to listen to Major General Carl E. Mundy III and CAPT Dana Gordon

PART TWO

EDUCATION

CHAPTER NINE –

Accept the Changes and the Challenges

We don't grow if we don't accept change and challenge ourselves.

You must accept those challenges and changes in order to become successful.

– Dana R. Gordon

After 34 years of affiliation with the world's greatest Navy–five years as an NROTC student and 29 years on active duty–I decided to retire. I took the advice from some former retirees and took 90 days off, which was the absolute best thing I could have ever done before I started my second career. I would highly recommend that you do this if you are able when looking to start a second career. It gave me time to decompress and change my mindset, to do many of the things I had been wanting to do personally, but most importantly it allowed me to catch up on the all important "honey-do list" that my wife and family had for me to do for them around the house. The break allowed me to totally remove myself from what had been an incredible journey, one that I had no idea

I would be taking as a then 18-year-old leaving my parents' home in Columbia, SC, and now I could absolutely exhale from a lifestyle that filled me with some of the toughest challenges and changes anyone could ever have imagined.

Through my first seven years of completing one of the most difficult college curriculums in America as an electrical engineering co-op student at Georgia Tech; to pledging and becoming a part of the best fraternity in the world, Omega Psi Phi Fraternity, Inc; to receiving my Officer commission, completing Naval flight school and then becoming a Naval aviator, I had no idea how those huge, life-changing accomplishments would help shape the next 27 years of my life and the level of success and achievements they would bring my way.

I am absolutely proud of every challenge that came my way throughout both my personal life and in my Naval career, and I can definitely say that it was those major milestones that helped shape my thinking and learning to propel me forward to complete some of my most difficult decisions and assignments. Those very unique challenges–educationally, physically, and mentally–helped change my thinking in so many ways that I can never put them all on paper. Suffice it to say, it was those challenges and changes to my lifestyle and thinking that helped to make me successful in the many endeavors I was able to achieve both in the Navy and now beyond them as I work in the private sector.

Had I never accepted each of those challenges head on and continued to grow from them, through successes and some setbacks, I know personally that there was likely a smaller piece that I may have missed later in life to help me in making decisions, be it on a military mission, during a personnel situation, or even later in my career in engineering or financial matters. Achieving those milestones and many others that I built upon weren't easy, as nothing worth having ever is, but don't be afraid to accept the challenges and changes that come with it, as you'll definitely find out how insightful and knowledgeable those experiences will become as you grow.

Georgia Tech Omega Psi Phi fraternity line brothers FourEver and a day: Walter Reed, Dana Gordon, Barian Woodward, and Anthony Moore

Young ENS Dana Gordon starting out his fledgling career as a Naval aviator

CHAPTER TEN –
CONSTANT COURSE CORRECTIONS

I never lose, either I win or learn.

– NELSON MANDELA

In life, winning and losing will both happen.
What is never acceptable is quitting.

– EARVIN "MAGIC" JOHNSON

Upon retiring from the Navy, I went through my interview process, searching for jobs online, discussing things with colleagues, and writing resumes and doing interviews before I found my current job as Director of Engineering at Fanatics, the worldwide leader in licensed sports merchandise. During my one-on-one personal interviews, I was asked some very challenging questions that I felt certain I had answered appropriately and confidently. After all, I had just left the Navy after 29 years having directed operations that saved thousands of lives and equipment through five hurricane evacuations, commanded the Navy's second largest vessel with 3,250 Sailors and Marines, successfully commanded a squadron, flown 3,540 flight hours and saved six lives during two very difficult helicopter rescues. Through six deployments, workups, and geo bachelor time that amounted to 11 total years of family separation, I still had my lovely bride of then 26 years, and my girls were doing very well as student-athletes.

What could they possibly ask me during an interview that, through life's experiences and the challenges and changes that I had overcome, that I didn't have an answer for? Then I got a very simple yet profound question from Skip Smith, one of the Vice Presidents at Fanatics who I would be working with on our team, that changed my thought process and put all of those achievements into perspective for me. "Tell me about a time you failed at something and what you did to correct it?"

It was a question that in retrospect, in preparing for my interviews, I probably should have expected and prepared more for, but truthfully, it caught me off guard. However, coming from the positions I had held and going to work as a retired military person now in the civilian world, it was the perfect question to ask. It was by far the toughest question that was presented to me during any of my interviews. Why? Because I personally had never viewed anything I had done as failing. Was this something I should take personally and say I failed to do something? Did I somehow let my Sailors, Marines, or teammates down because I failed to do something? I think the word *failed* hit me more than anything else, and so I thought about it for what seemed like a full minute or two before I provided an answer.

That answer was, "I've had numerous setbacks where things didn't go as planned, but I wouldn't call them failures. There have been times where we didn't meet our objectives completely but were still able to adjust as we went along, learning something as we eventually got to our ultimate goal. If somehow, it wasn't achievable and it was ultimately deemed a failure, then we looked at why we didn't get there, and we tried to learn what stopped us from getting there."

Now to some of you this may sound like some glass half-full BS answer, and you may say yea, he spent two tours in DC at the Pentagon, so he knows how to lay it on thick, but the truth is, we rarely ever achieve 100 percent success at anything we set out to do. You see, when we set out on a task or a mission, we are constantly learning and adjusting as we go about trying to be successful, be it personally or while leading our groups or teams. Sure I probably may have sounded a little arrogant, but I just don't think I could or would ever call the work, plans, and efforts of what I or those who have worked for me failures if we didn't ultimately achieve our initial desired objective.

We had setbacks or fell short of that 100 percent goal, and I as a leader and planner definitely learned something from it, as did my team or teammates, but we didn't fail. In most instances, and I think this will be true in your life experiences as well, you'll have more wins than you will find yourself falling short, but if you do, always treat it as an opportunity to learn. You'll find that you and your team will become better because of it, and the next time will look to think through those events that made you fall short of your last goal.

In the end, I don't think you'll ever really "lose or fail," but you'll definitely win at some level and learn something from each and every experience you and your teams work through.

Great Team, Great Teammates, Lifelong Friends,
Lessons that last a Lifetime
Columbia High School Basketball
Conference and Lower State Champions
Class 3A State Championship Runners-Up

CHAPTER ELEVEN –

YOU'RE ALWAYS COMPETING

Every morning in Africa, when a gazelle wakes up, it knows it must run faster than the fastest lion or it will be killed. Every morning when a lion wakes up, it knows it must outrun the slowest gazelle or it will starve to death. It doesn't matter whether you are a lion or a gazelle, when the sun comes up you had better be running.

– AFRICAN PROVERB

It's a competition, not an exhibition.

– DANA R. GORDON

At Georgia Tech we would have lectures in one of the large physics and calculus lecture halls that could seat about 500 students at a time. During the first few days on campus, before the quarter started, they cycled through 450 of our then approximately 2,500 incoming freshmen into the lecture hall to complete what was our student in-doctrination courses. They did this over the course of a few days, and by the time they were done, we were well on our way to becoming graduates from one of the most prestigious engineering institutions in the world, the Georgia Institute of Technology.

Before we left that final indoctrination lecture and were sent on our merry way, however, we were all given a small dose of reality from the admissions counselor which went like this: "Look to your left ... look to your right ... only one out of three of you will be sitting here after your freshman year." Well, when you come in as the hot shot, Top 10 high school graduate from your class, you typically just laugh those types of things off–until you actually see it happen to others, and the threat to you personally becomes real!

Those words rang true, and through what were tough freshman and sophomore years, I was able to maneuver my way through what we called the "Weed Out" courses and eventually made my way across the stage where I proudly received my electrical engineering degree. However, there were some serious life lessons learned during my time there that taught me true discipline and brought out a level of educational competitiveness that before then I hadn't had challenged.

What I learned in the end was that sometimes you're competing against other people; sometimes you're competing against the standard; and most often you're competing against yourself, but what you should never forget is that *you're always competing*. If you want to be successful, and you're competing to be the best at a task against a group or even in some cases an individual, you should never lose sight of that fact, for when you do, you'll find yourself on the outside looking in as possibly two of the three that don't make the cut.

Life ain't an exhibition where you get to show how well you can do things; it's a competition based on performance in the job or task you have been given. So if you want to compete, when you wake up, you better be running…

The sun is up…You better be running!
The lion vs the gazelle

Beating out the competition, Georgia Tech graduation
Dolores, Dana, and Bobby Gordon

CHAPTER TWELVE – FIND THE LONG POLE IN THE TENT

It's always the French fries.

– DANA R. GORDON

Studying queuing theory in an Ops Research class while working on my master's degree in Business Administration, our professor asked us a simple question, "What is it that causes the line to back up at a fast food restaurant's drive-thru?"

Now having had this happen to me several times throughout my childhood and early adult life, I knew right away what the answer was, but was somewhat amused at the answers that came from my fellow classmates. Inexperienced cashiers, cash vs debit cards, and a mix up of orders were some of the answers that were given, and while they were all pretty logical and made sense, I listened to them all and chuckled as I waited to give my answer.

The professor could see me continuing to smile and shake my head every time an answer was given. Finally he called on me and said, "Mr. Gordon, you seem to think you have the right answer?"

I laughed and said proudly to him, "Why, it's the French fries, of course." Now much to my surprise, that wasn't the answer he had come up with as a part of this lecture, but he was quite amused at the answer I had given. So much so that he asked me to expound on why I so firmly believed my answer was the correct one.

I said simply, "Because every meal that you buy has French fries in them, and once they run out of fries, they will come to a grinding halt until they can put more fries in each and every bag that goes out the door, whether it's a walk-in or drive-thru."

He asked me what scientific or mathematical evidence I had to prove this, so I turned and asked two simple questions of my classmates: 1) "When you go to get fast food, how many of you get fries?" All hands went up in the affirmative, and 2) "If you have to wait, what's the primary reason you are given?" All agreed it was the fries!

The professor laughed and stated that in his many years of teaching the class, he'd never looked at the drive-thru problem from that aspect of things. His mindset was focused on the multi-tasking required of the employees and also, at that time (1997), the lack of automation in the order processing. He never thought of the basic fact that the one common element to almost every order that went out the door at fast food restaurants was their French fries. We all laughed about it that day, and "It's the French fries" was the answer to every hard question the rest of that semester, but I never forgot that story and the lesson I learned that day. From that point on, I always looked at things differently when trying to solve problems. I looked at taking the viewpoint of trying to find the common denominator or common thread in a situation to see if by improving that, it could possibly solve multiple problems.

Some people call it the "long pole in the tent" in reference to the old circus Big Top tents from back in the day. You see, without getting the longest, biggest pole in the circus tent put up first, the show never starts because everything revolves around the tent being up and staying steady around that long pole. Some leaders never understand how to solve problems by looking at the "long pole in the tent" and often find themselves fixing smaller, multiple pieces to a much larger problem. I found by listening to and taking in all those problems and trying to find a common thread to solve for, you can often solve the majority of them all at once.

I'm not saying this method will solve all your problems, but often if you can fix the "French fry" drive-thru dilemma by determining what element is the common denominator that holds things up, you'll probably end up solving most of your issues.

CHAPTER THIRTEEN -

You're Only as Smart as Your Glasses

I have not failed. I've just found 10,000 ways that won't work.

– Thomas A. Edison

If my mind can conceive it, and my heart can believe it,
then I can achieve it.

– Muhammad Ali

You're only as smart as your glasses...
If you don't see it and learn it,
you won't be smart enough to teach, lead, and influence it.

– Dana R. Gordon

There have been millions of inventions created in our lifetimes. From the lightbulb, to television, to the computer, the internet, and cell phones, there are things that have forever fundamentally changed the life and lifestyles of man and how we live.

I always smiled when someone who I worked for would say, "Let's find a solution here and 'think out of the box'..." Early on in life, especially as a young leader challenged with making a difference, I always wondered what exactly that meant. I mean isn't doing something outside of the norm often frowned upon when you are in a working environment, and exactly how far "out of the box" should I attempt to go when making a decision or leading my team? And more importantly, how far will my leaders allow me to "get out of the box" before they reel me back in?

As the world becomes more complicated and things become more dependent on other influences, our lives, decisions, and yes, even our solutions to problems become more and more complex. Sometimes you can work the trial and error method as Edison would suggest, making tens of thousands of attempts before he was successful at refining the incandescent light bulb. I would say today, however, if you were given more than a handful of attempts at success in anything, that short of being an inventor, you would be out of a job or would get fired if you were a coach.

Even "The Greatest," one of my all-time heroes, Muhammad Ali, would suggest that with the right imagination, conceptualization, and determination, just about anything is possible. This is often the case when we see miraculous feats achieved by some incredible athletes, heroic, often lifesaving deeds performed by men and women, and bright ideas that come to life as a result of an idea that someone has that will "make a difference."

I personally have never fancied myself the inventive type like Edison nor was I blessed with the ability to "float like a butterfly and sting like a bee." I would say however, that there is probably a way you can figure out how you can work towards being one of those "think out of the box" type leaders and make a difference in leading your team to success.

I have always found that LBWA, Leadership By Walking Around, was the best way to come up with solutions to the majority of the problems needed in the organizations I was a part of. I found that until you took the time to learn the processes, procedures, and most importantly, the people skills required to do the job, by seeing it firsthand, you could not and should not attempt to provide input to improve a process. Until you know first hand, what it truly takes to get something accomplished through your own educational insight, you will not be smart enough to teach, lead, influence, or speak on behalf of those who are performing that task.

As leaders we are charged with helping to make improvements in our daily processes. Some are to make money, some are to save money, others are to protect or even save lives, but all of them are key decisions that matter and if done correctly are game changers in what we do daily. Know that if you aren't fully immersed and knowledgeable in the task you are trying to improve then you will never have a complete understanding of how to truly make things better or more productive.

I am so proud of the military training that I received both in the world of aviation as well as onboard our Navy ships. I was trained by the absolute best junior and senior enlisted Sailors and officers in the details of how to do business. We were trained as if our lives depended on it because they literally did. By following their lead and learning daily from my Plane Captains, who prepared my helicopter before I went flying, I learned the nuances of what to look for that could be dangerous and out of tolerance. By performing spot checks and making daily rounds with the Sailors onboard my ship, I was able to learn what was important in the operations of their equipment, and more importantly, what they personally worried about in helping to keep their equipment running properly.

Putting eyes on equipment and operations as a leader is definitely important, but having conversations with those same operators is an absolute must. It's this last step in the educational process that takes it from your visual learning to the ability for you to make decisions that affect improvements that are game changing. It allows you to think out of the box without affecting the operators' ability to perform their jobs, because you fully understand how the decisions you make may affect their performance because you've been there talking with them before.

In my current job here at Fanatics, I've had the opportunity to put many of these same practices in place as I visit the six facilities that my team and I are responsible for. Our goal is to make continuous improvements from the engineering and materials side of the house to increase operational performance or provide costs savings in some way. It is this same emphasis on hands on learning of the processes, utilizing that same LBWA, that I learned in the military, that we all practice as well at Fanatics that has allowed us to make significant improvements in our processes throughout the company that all of us collectively are truly proud of.

So you don't necessarily have to be one of the most famous inventors the world has ever known or the greatest athlete of all time to make creative and inventive changes and differences. Each of us holds that ability and can gain that knowledge by learning how to be "as smart as our glasses," and getting out and not just seeing, but truly understanding the processes from every aspect, since as we all know, "a picture is worth a thousand words..."

CAPT Dana Gordon, Change of Command ceremony, USS IWO JIMA

CHAPTER FOURTEEN –

Choose Your Hard

Don't expect people to understand your grind when God didn't give them your vision.

– Russell Wilson

Marriage is hard. Divorce is hard. Choose your hard. Obesity is hard. Being fit is hard. Choose your hard. Being in debt is hard. Being financially disciplined is hard. Choose your hard. Communicating is hard. Not communicating is hard. Choose your hard. Life will never be easy. It will always be hard. But we can choose our hard. Pick wisely.

– Devon Brough

Driven, goal oriented, focused… What term do you want to be labeled with? I'm sure any of those three are great to have on your resume and be associated with if someone talks about you and your work ethic. So why is it that some people achieve heights that make them the G.O.A.T., while others never quite get there?

Nature vs Nurture… Everyone is an individual with their own thoughts, ideas, and goals; yet even though they can be brought up in the exact same environment, they can end up going down two very different paths and having varying levels of success in their lives. While Nurture is the standard that we all live by based on who raises us, it's the inherent Nature within us that determines what "grind" we choose and how "hard" we're going to work to get there. It's always wonderful to see those who are hard-working and grinding to get better, as I'm a firm believer in hard work. However, I personally have always been of the mindset that you should work "smarter not harder," and therefore have always looked for the more efficient way to do things. I suppose that's the true engineer in me and why I love solving problems to make things more efficient for myself and others.

What I've also come to grasp and fully appreciate is combining that more efficient process with a solid, hard-core grind to get something accomplished. Staying up late and "choosing my hard" when it matters and giving the extra effort to make sure it's right the first time is what I always strive to do to get the best product possible, both personally and professionally. All of those decisions are inherent in each individual, and as Russell Wilson points out, are part of the vision and desire that have been given to each of us when we set out to achieve a specific goal or to obtain an objective.

We all have many choices in life so "choose your hard." Set your goals, both personally and professionally, and figure out what you want to go after to be successful. It's up to you to determine how far you will go and how much you will achieve, but all of that is based on your vision, your hopes and dreams coupled with how hard you want to work to attain them. Pick your vision, set your goals, and choose your hard to determine your level of success.

Retirement ceremony, Navy family departing CAPT and Mrs. Dana Gordon and family being piped ashore

CHAPTER FIFTEEN –

HONESTY AND RESPECT FOR OTHERS

Honesty and transparency make you vulnerable;
be honest and transparent anyway.

– MOTHER THERESA

Best way to earn respect is by treating others with respect.

– DANA R. GORDON

Everyone you will ever meet knows something you don't.

– BILL NYE, SCIENCE GUY

I was asked to participate in a Black History month program where I had the opportunity to talk and interact in a Zoom session with over 300 brilliant STEM high school students. I was given a few questions ahead of time and then answered a few more to close out the hour-long session.

One of the questions I received was, "Throughout all the different occupations you have had, what was your primary guiding principle?" I thought it was one of the best questions I had been given all day, as it definitely made me reflect on what basic tenants had guided me both then and now throughout my life and career. I settled on two basic things: honesty and respect for others. In my early career in the Navy I realized how much honesty and integrity played a role in our daily lives and was the very fabric of our missions.

As a pilot, our very lives depend on those who work on our aircraft: our maintenance professionals and the plane captains who ultimately sign my aircraft safe for flight. Often these responsibilities are given to some of the youngest members of our squadrons, and we as pilots trust that they will do their job as per the Standard Operating Procedures. We believe in their honesty and trust their word when they sign the final piece of paper that says our aircraft is, "safe for flight." Our very lives depend on the honesty of each of those magnificent Sailors being honest, trustworthy, and doing their jobs correctly and accurately.

That trust and honesty doesn't stop in the aviation world. It is also a part of our ships, on our submarines, and indeed is the very fabric of every branch of all of our military services, for if you cannot trust those who are a part of your team, you and the team are doomed to fail.

I have this same level of trust in my current occupation with my fellow co-workers, my engineers and other members within our Fantastic Fanatics Team. Knowing that each of them will "do their part" ensures that we will be successful as a company, just as I was successful in the many missions I was a part of while flying or commanding while onboard my ship. I also believe respect is interwoven in everything you do in life. I am a firm believer in the basic tenant that you have to give respect to get respect.

Everyone I have ever met is a fascinating individual with a unique story to tell. Their lives, their experiences, their trials and tribulations, successes and failures are what make them who they are. I have always found myself wanting to learn about the person before I ever dive deeper into the qualifications, achievements, and accomplishments they may have.

While we don't realize it, the simple fact that everyone knows something you don't is the basic tenant behind the beauty of diversification and strengthens how we interact with each other. The ability to learn something I didn't know always fascinates my curious mind and keeps me engaged with those who are currently a part of my circle or who may be new to our group.

Getting to know who that person is plays a large part in establishing a mutual respect for each other and is also a key component to making your team successful. I am a firm believer that without both Honesty in how you operate and a Mutual Respect for others, you cannot be a successful team and win at any level or in any profession.

Relieving of The Watch

Led by Command Master Chief Mack Ellis with mentee and former junior enlisted mechanic and Plane Captain, now LCDR, Roy Gugia and his son A1C Ioannis Gugia, USAF, relieve CAPT Gordon of his duties one final time at his retirement ceremony.

CHAPTER SIXTEEN –

NEW DOESN'T MEAN BETTER

It's like putting a bad pilot in a new aircraft; the new aircraft doesn't make the pilot or mission any better if the pilot still doesn't know how to do their job.

– DANA R. GORDON

This was a joking statement I made one day when talking to one of our facility General Managers, Retired Navy Admiral Ron "Beav" Horton, former F/A-18 jet pilot and aircraft carrier CO. It is one that I have always found to be true when discussing automation and/or process improvements no matter where I have been, both in the military or in the civilian world. Sometimes we all fall into the trap of thinking "new" means "better," or that it will automatically fix the problems or issues we face. Many times however, there are other underlying problems that need to be improved upon through training and education before the enhancement, automation, or upgrades are introduced. Otherwise, we open Pandora's Box and out pop a host of other unintended consequences that we may not have foreseen ahead of the improvements we're looking to introduce.

I'm an engineer and also a pilot and so bringing those two worlds together for me has shown me through my past that as an operator, you must know the "mission" or your stated goals before you seek to ask for upgrades or improvements in your processes. If you have a solid knowledge of what you are trying to achieve and know the deficiencies you have that are barriers, then you are more likely to have an appropriate ask for a solution, rather than taking what may be presented to you at face value as the answer.

Many times we have seen the "next best thing" come along to solve the world's problems for someone, only to watch those who get it struggle to get their desired results. You end up installing or implementing something that may get you partial results, which then cause you to expend more energy and resources in another area that you hadn't thought you would need to. Hopefully the ones making the decisions to bring forward those new ideas are held to a standard of providing a return on investment or demonstrating the ability for the process to become a force multiplier based on the solution set they are recommending. Without either of these, coupled with a lack of understandable shortcomings, any project, new piece of equipment, or new process may end up creating more problems than you may have been ultimately trying to solve.

Bottom line on this one is to ensure you know your ask and if you are the one attempting to solve the problem, tailor the solution for that ask. Don't make the mistake of providing a solution that is looking for a problem to solve and creating additional problems that actually don't fix the current situation.

HSL-42 Proud Warriors Squadron Formation flight off the coast of Mayport, FL Commanding Officer CDR Dana Gordon in lead Squadron Painted Aircraft, with CDR Chris Failla, XO in Dash Two followed closely by some of the world's finest Naval pilots and aircrew in Dash Three through Seven

PART THREE
MENTORSHIP

CHAPTER SEVENTEEN –

BE A BRIDGEBUILDER

It's nice to be first, but it loses meaning if there is never a second. Build that bridge and keep pressing to make a difference.

– DANA R. GORDON

This very prophetic statement is one that has driven me throughout most of my life and career. My very proud mother, a 32-year middle school, Special Education teacher, was always one to "suggest" and ensure that I try to be the "first" at something, especially when it came to being a young Black male. She always gave me encouragement while looking for a path to discovery to do so whenever an opportunity presented itself.

As I began to achieve things, she would start to highlight and remind me (and many of her closest friends) of those accomplishments as the first Black male to do so: first Black male senior class president at my high school, first to achieve a coveted, then full $80K Navy ROTC scholarship at my high school, first Black male to attend and successfully complete

Navy flight school from the GA Tech NROTC program, first Black officer at every paygrade from Ensign to Commander in each of my Aviation Squadrons, first department head and Officer in Charge in my Squadrons and also onboard ships I deployed in, first Executive and Commanding Officer to serve in a command role of both my Squadron and my ship, first Aviation Diversity Officer to lead the Naval Aviation Enterprise, and first Black Chief of Staff for the Flag Officer at Navy Region Southeast.

But while I relished all those accomplishments and was absolutely proud of each and every one of them, I realized two very important things: I never got there without the strong support, guidance, and mentorship of others, and, most importantly, none of my "firsts" meant anything if there wasn't a second young and aspiring Dana Gordon coming behind me to do the same or even better.

Having what I call a third generation school teacher's mindset, with a grandmother, mother, and sister all being educators, I have often said that I have teaching and mentoring in my blood since that's the type of environment I was raised in during all of my formative years. Through many years of observing and from those teachings, it was instilled in me to always look to do better but also remember to give back and "pay it forward" as much as I could. It is because of those values that I have never shied away from being a mentor and role model to whomever I could.

While mentoring can happen formally or informally and in numerous ways, I have often found that it is always the most successful if you first establish a solid understanding of the personality, goals, and aspirations of the person with whom you are working. I have proudly mentored both men and women from so many diverse backgrounds that there are far too many to mention here, but the one thing I am proudest of is that through each of those wonderful relationships, I can safely say that I have gotten as much, if not more, from each of them than I could have ever hoped to give in return. From Sailors like Elliott Youngblood, Prince Adu Darko, Catherine Stevens, and Jeremy Bartowitz to Roy Kindolo Gugia, Donte' Jackson, Captains Dewon Booker, and Dave Loo, Colonel Merryl David Tengesdal, and Admiral Stephen Barnett, I am so proud to have had the opportunity to have discussions with each of them and to help contribute in some small way to any success that each of them may have had during their wonderful careers and/or personal lives both inside and outside of the military.

More importantly, however, I am just as grateful for the experiences each of them has given, in allowing me the unique opportunity to learn and grow personally with each of them. Seeing their success and the success of the thousands of other persons I may have possibly influenced along the way is what motivated me then and continues to motivate me still today, as the mentoring process never truly stops.

Thanks to all those who allowed me the opportunity to "bend their ear." I appreciate you listening, but more importantly, I appreciate you providing me feedback as well to help me in my own personal growth and development along the way, as I know I have learned as much from you as you all may think you have learned from me. It is because of this education and learning experience that I am thankful to each and every one of you.

Celebrating retirement ceremony of mentee and close friend LCDR Elliott Youngblood and CAPT Dana Gordon

"Old Glory" Retirement Flag Passing Ceremony

CAPT Dana Gordon, CAPT DeWuan Booker, and CDR Donte Jackson

"Old Glory" Retirement Ceremony Final Flag Salute

Looking back at my history, but forward to the future... CAPT Dana Gordon, Command Master Chief Mack Ellis, CAPT DeWuan Booker, CDR Donte Jackson, LCDR Hector Ferrell, LT Clifton Johnson, LTJG Brandon Russell, ENS Michael Pullium

CHAPTER EIGHTEEN

Put Me in the Game, Coach!

You should always be given an Equal Opportunity…
An Equal Opportunity to fail or succeed.
If you haven't been given that chance to perform
and prove yourself, then you haven't been given
an Equal Opportunity.

– Dwight Shepherd, Rear Admiral,
USN (Ret)

In the world of sports, if you truly want to win, it's not about being fair, but about who's the best player you have that gives you the chance to win that should determine who you put in the game.

In the game of life, the question is always: will the coach (your boss) put you in if you're the best? Will they give you the chance to shine and make sure others know you're the best by giving you that credit if you're successful? Will they be your champion and your mentor? "You play to win the game!" was a famous quote by Coach Herman Edwards when discussing why you participate in sports; just like in sports, everything in life is the same way. If there is an objective or goal, you should be playing to be the best or to win.

I have been so very fortunate in my lifetime to have had mentors who "put me in the game," who allowed me to grow by both failing and succeeding, but also by providing me with guidance and great mentorship while championing my cause along the way. From all of my teachers, to my many coaches in Pop Warner football, through middle and high school basketball, to my numerous military mentors and bosses along the way, to the current leadership team at Fanatics Inc., I am so grateful to have had so many outstanding mentors who have given me an Equal Opportunity (to fail or succeed) at so many tasks that I have been given.

As a boss or coach, you should always be ready and willing to mentor and seek out opportunities to challenge those who work or play for you, be it in a working environment or on a field or court. Being a mentor isn't easy and requires time, effort, and a lot of patience, but the reward you receive by watching your mentees grow and succeed has always been the most rewarding thing I could ever achieve as a leader.

Winners! Naval aviation and helicopter community's best and brightest CDR Alvin Holsey, CDR Dana Gordon, and CDR Gary Mayes

CHAPTER NINETEEN -

WHO'S YOUR DADDY...?

In the Navy we usually have strong mentors that we call Sea Daddies (or Mamas) who lead us through the ups and downs and teach us the many facets of life as a Sailor. The good ones will often have those frank conversations with you and give you the "behind the scenes" discussions necessary to help you navigate through your entire career and make you successful.

During my time in the Navy, I was blessed to have two Sea Daddies, Rear Admiral Earl Gay, USN (Ret) and Retired Navy Captain, now Mr. Neil Hogg, SES, who grabbed me at the very beginning of my career and still both guide and teach me valuable life lessons to this day. Between the two of them, I was able to learn every aspect of the Navy, be it managing life on deployments, to flying an aircraft, commanding both a squadron and a ship, to the business and gamesmanship that is involved with surviving, promoting, and becoming a solid leader of Sailors and Marines. I could not have been more fortunate than to have had two of our Navy's finest leaders decide to take me under their wing and provide me with the guidance and insight to make me as successful as I was in my 29-year Naval career.

One thing that I think helped me the most in my growth and development was the fact that they are absolute polar opposites in almost every way, shape, and form. From personality, to race, and even in leadership approaches to some things, they are definitely two very different people. For me, it was an absolutely ideal situation because I could get the best of both worlds since I knew that they both had my best interests in mind and that I would get nothing but the "straight facts" from them both without it being sugar-coated–something that you should always get from a true mentor. Their mentorship allowed me the opportunity to learn how to manage and deal with situations from two very different viewpoints and formulate my own decisions both in leadership and also in my personal life throughout my career.

I will tell you that I certainly would not have been nearly as successful and been able to lead as effectively without the sound advice and judgement I received from both of them, sometimes when it was solicited, but often when it was unsolicited. In either case, I was more than happy to receive and take onboard whatever nugget of info was passed my way to help me progress.

As a mentee, looking to find a mentor, I would recommend that no matter what profession you are in, you should look for at least two mentors in your profession that are not too similar. The diversity of thought, opinions, and backgrounds by each will provide you with invaluable insight from very differing viewpoints and will absolutely broaden your scope on life and make you a better leader.

I would often tell my Sailors that you should find a mentor who looks like you and one who doesn't–one majority and one minority. If you are a female, you should definitely have at least one male and one female mentor at a minimum. Hearing the differing truths, life experiences, and insight from both perspectives can only make you a better leader in the future.

Without them, there would be no me...
Mentors who led the way
Mr. Neil Hogg, SES, CDR Dana Gordon,
RADM Earl Gay

CHAPTER TWENTY –

MAKE A DECISION, BUT DON'T TAKE TOO LONG

When you come to a fork in the road, take it,
but make damn sure you've done your research first.

– DANA R. GORDON

I love the beginning of this quote by the great Yogi Berra. It's definitely one that supports my theory of not standing still or waiting to make a decision. The rest of it came from me after a discussion I had with one of my former COs, CAPT Steve Bagby, who once told me, "Dana, we're gonna make a decision and we're not gonna take too long to do it, so what's your thoughts on where we need to go?"

The first time this was presented to me by Skipper Bagby, I didn't have an immediate answer, so he told me to come back after doing my research, come up with a few possible answers, and we would make a decision together on the solution. That small piece of guidance, mentorship, and leadership has been one of the fundamental constructs by which I have lead and mentored throughout my entire career. From that point on and for every person I've worked for ever since, I've always come to them with at least two, sometimes three, solutions to a problem, and I have asked that those who work for me do the same.

No leader wants to solve all the problems you bring to them, nor do they have time to do so. With all the complex things I personally had to manage throughout my career both during and now after the military, I have always been absolutely appreciative of the numerous leaders and managers who presented me with issues or problems, along with multiple ways to solve them. Because they took the time to do their research before presenting me with the problem, they already thought through the "what ifs" and could often answer my questions when I asked them. It challenged them to think through the problem and also made both them and those who worked for them that much smarter, so we often could come to the correct decision the very first time, without taking a misstep. We saw that fork in the road, we made our decision, and with solid research in hand we took it!

CHAPTER TWENTY-ONE – THE MOVING FINGER WRITES

The Moving Finger writes; and, having writ, Moves On: nor all thy Piety nor Wit, Shall lure it back to cancel half a Line, Nor all thy Tears wash out a Word of it.

– OMAR KHAYYAM

Those who fail to learn from history are condemned to repeat it.

– WINSTON CHURCHILL

If you could go back and change one thing in your life, what would it be?

I've often been presented with this question by many who I've talked with, by those who I have mentored, and from those who have worked for me, and my answer has been quick and very simple…

Nothing.

The past and those wonderful experiences we've all had, good and bad, are what make us who we are and shape us into who we will become in the future. Those actions we have taken in the past are the experiences that will drive and influence us in our decisions as spouses and parents, as leaders, and sometimes cautiously, as followers of others into the future.

Sure I've made mistakes in life; we all have, but the worst thing you can do is to make a mistake, not admit it, and never learn from it. I have, however, never been a proponent of trying to go back and change the course of history by changing how something may have affected me in the past. Who knows how actions that were perceived positively or negatively at the time may have changed or influenced things for you in the future. Sure there are things that we'd love to take back and change in our lives, and some people have certainly had tragic events occur that probably should be considered, but the majority of the decisions we make in life are ones that we live with and ultimately learn from. Based on these thoughts, and holding to the insightful words of Churchill, I would hold firm on my decision to continue learning from the past and use those experiences and insights to help me make smart, insightful decisions for the future.

CHAPTER TWENTY-TWO – HAVING PATIENCE WILL PREVENT REGRET

A moment of patience in a moment of anger,
prevents a thousand moments of regret.

– ALI IBN ABI TALIB

One of the great leaders I had the pleasure of working with during my command tour onboard IWO JIMA was my Legalman, Senior Chief Petty Officer, Patrice Kelly. She was the ship's legal conscience during my tour as Executive Officer and for a short time for me while I was the Commanding Officer, where she managed every legal matter we had onboard IWO.

Now you can imagine being on an amphibious assault carrier with 3,250 Sailors and Marines for workups and an 8-month deployment, that there are bound to be a few rules broken and that something would need to be done to ensure good order and discipline were maintained onboard. Often a steady hand and solid background knowledge of the legal facts are needed when things are presented to you that would otherwise just plain set you off. LNCS Kelly was that steadying influence for both my previous Captain Jim McGovern and ultimately myself as Commanding Officer, as we waded through the often murky legal world that was the Uniform Code of Military Justice.

Now I've never professed to being a lawyer and wouldn't even proclaim to have stayed at a Holiday Inn to try and be one, so we leaned pretty heavily on her knowledge and expertise during all legal matters we encountered. Fortunately for us, it was always her patience and calm demeanor in group conversations behind closed doors with leadership that allowed us to come to the right decisions on the cases that would ultimately determine the fate of our Sailors and Marines. As my trusted legal advisor, it was always her guidance that allowed me to often maintain my patience during what could have been moments of anger and kept me from saying or doing many things I may have regretted later, possibly to the detriment of my career.

I have watched many, many leaders ignore the advice of their knowledgeable and trusted advisors, often leading to their own demise. While you may be in command or the positional head of some group or team, there are usually those subject matter experts who are put in your organization to advise you on how to best manage or make a decision in their particular area. They are put in place quite frankly, if for no other reason, than to keep you out of trouble.

As a leader, it is your responsibility to learn to listen to those advisors, take in their info and make the best of what you have in front of you before rendering or making a final decision on things. While you may not always agree with them, my advice to you is to always listen to what they have to say. If you disagree, be sure you've discussed the unintended consequences of not doing what they advise before going out on a limb to do so, because in many cases, once squeezed, you can't put that toothpaste back into the tube.

I was so grateful to have LNCS Kelly and her relief, LNC Carin Deitler, as my Legalmen during my time in command, as both saved me from thousands of moments of regret during many of my complicated decisions while serving onboard IWO JIMA. Thanks to you both for your patience and professionalism.

CHAPTER TWENTY-THREE –

LET THE LEADERS LEAD

To build a strong team you must see someone else's strength as a complement to your weakness, not a threat to your position of authority.

– CHRISTINE CAINE

Leaders don't create followers; they create more leaders.

– TOM PETERS

A great boss is hard to find, difficult to leave, and impossible to forget.

I always love when those who work for me succeed at something they do. As a leader, it should be second nature and woven into your DNA that you would want those who work for you to succeed and ultimately become as successful, if not even more successful, than you are.

I have had the absolute pleasure of working with some of the most talented and smartest people on this planet, both in and out of the military. As a leader, I believe it is a must that you seek to find the strengths in everyone who works for you and then utilize those strengths to make the team better as a whole. We all have our weaknesses, including us as leaders. If as a leader you think you have all the answers, then you are absolutely doomed to fail.

I always welcomed new challenges in my life and my career. I think it's what has kept me going all these years and gives me fulfillment. I never touched the controls of any type of aircraft before my first flight in a military aircraft and went on to complete over 3,540 mishap free flight hours and command a squadron of 14 helicopters. After 20 years of flying, I was promoted, selected, and completely switched careers in the Navy and learned steam engineering, shipboard handling, damage control, flight deck and well deck operations and became an amphibious assault aircraft carrier commanding officer.

Now mind you, both of those are somewhat extreme examples, but my success in both of those endeavors didn't come without a significant level of confidence in those who worked with and for me. I placed a large amount of trust in the professionals in their fields to guide me through the areas where they were the experts. While I learned how to fly an aircraft, there were none finer at making sure those aircraft flew and were maintained than SR Chief Don Warner and Master Chiefs Bill Lukander and Gary Donner. A ship doesn't run without it's engineers and without the efforts of CDR Charlie Lynch, CWO4 Hurdis D. Rogers, or CMDCM Max Mullinax,

USS IWO JIMA, The Magnificent Seven would have never left the pier.

These six tremendous leaders and their teams were the absolute best at what they did, and as a smart leader, I knew this and did what every good leader should always do when you realize this–*get the hell out of their way*– and let them do their jobs. As professionals, these magnificent leaders were just as driven to make both our squadron and ship as successful as I was, and so I knew that the success of their teams meant our overall success was inevitable.

One of the things I discovered as I worked with these exceptional leaders was that they left a legacy at every command they were a part of. They had an absolutely great following with Sailors and organizations that achieved outstanding things throughout the Navy, but more importantly, their longer lasting legacy was that they created some of the best leaders that are still in charge of squadrons and ships throughout our Navy today. It's phenomenal bosses like Fanatics Chief Customer Officer Lonnie Phillips and Continuous Improvement Vice President Joe Matthews, along with great military leaders like Admiral Kevin Delaney, Captains Glenn "Popeye" Doyle, Wayne "Great Tutini" Tunick, John "Blast" Furness, Steve "Bags" Bagby, and Jim McGovern who you never forget. Their leadership styles, all very different for me, have given me tidbits that I have collected along the way and put into my kit bag of leadership tools to help me manage the daily leadership challenges I may face.

I'm sure you all have had great bosses in your past; be sure you capture those same tidbits and pearls of wisdom that they each impart to you as you grow and cultivate your own set of leadership tools for the future.

Best CPO Mess in the Fleet

Chief Petty Officers from USS IWO JIMA led by Command Master Chief Max Mullinax

BEST OF THE BEST

HSL-42 Battle Efficiency Award presentation
Commodore Glenn Doyle and Squadron
CO, CDR Dana Gordon

CHAPTER TWENTY-FOUR –

IT'S OK TO BE UNCOMFORTABLE

Be willing to be uncomfortable.
Be comfortable being uncomfortable.

– PETER MCWILLIAMS

A true testament to growth and development as a person, leader, or mentor is your ability to learn how to get comfortable being uncomfortable.

There are a great many challenges we are presented with both personally and professionally: shooting free throws during a close game, public speaking, taking a penalty kick in a soccer match, asking questions in a group setting, leading a group for the first time, or even speaking out when an obvious wrong has been committed. These are but a few examples. What I have found is that through education, repetition, and eventually a level of confidence, all things uncomfortable begin to become comfortable. And while you never quite fully become the "expert" at the thing you are doing or have been uncomfortable with in the past, you tend to gain an appreciation for it and learn to have a healthy respect for what you are ultimately trying to achieve.

During my many years in the Navy, through the trials and tribulations of life and dealing with the lives of others, I have come across a few situations, events we often call "Sea Stories." I'd say that dealing with those events, and sometimes the individuals involved, has put me in some quite uncomfortable situations that I've had to work my way through. They challenged my decision making, my thought processes, and in many cases tugged on me personally, as sometimes the decisions I made meant Sailors would be released from the Navy, which would have a significant impact on not only the Sailor, but also the family members they supported and the very lifestyles they were currently living.

Because of this, I would often lean on the sage advice of my mentors, peers, or even my Master Chiefs, Senior Enlisted Advisors, or other subject matter experts to help me in making those very tough decisions. One thing I came to realize during each event was that each and every decision was different with different variables, and depending on the solution, would have a different impact and outcome, so I took each one very seriously. However, eventually I became comfortable with making these sometimes uncomfortable decisions.

How? Through preparation, by gaining as much knowledge as I could about the individual, the situation, or the circumstances, so I was as well informed as I could be. Secondly, as I noted earlier, I asked for advice to make sure I wasn't off base in my thoughts or decisions. Through the beauty of having a very diverse group who worked for me, I would often discuss the matter with those who I knew might have differing viewpoints or opinions regarding the situation so I could see things from every angle.

It was through these various lenses that I was able to help work through some of my most uncomfortable decisions. Now I'm not saying I got them all right, but knowing that I had done my due diligence in working towards a solution gave me the comfort level in working through many uncomfortable situations. In my discussions with those I mentor, I often talk about the dilemma of dealing with being uncomfortable since as a leader, I think it's perhaps the most challenging thing we face. Learning, growing, and thoughtfully working through those challenging circumstances and decisions are what, in the end, make us better leaders, mentors, and role models for those who we work with and who work for us.

One of the greatest challenges,
but most rewarding experiences of my life…

CAPT DANA GORDON,
THEN XO USS IWO JIMA